BOEING 737

P. R. Smith

Copyright © Jane's Publishing Company Limited 1986
First published in the United Kingdom in 1986 by
Jane's Publishing Company Limited
238 City Road, London EC1V 2PU
in conjunction with DPR Marketing and Sales
37 Heath Road, Twickenham, Middlesex TW1 4AW

ISBN 0 7106 0424 6

Printed in the United Kingdom by Netherwood Dalton & Co Ltd

JANE'S

Cover illustrations

Front: **Aer Lingus – Irish Airlines** (EI)
Aer Lingus was formed on 22 May 1936 with the assistance of
Blackpool and West Coast Air Services, operations commencing
from Dublin across the Irish Sea to Bristol using a de Havilland
Dragon. Today the airline flies a scheduled passenger and cargo
network to major points in Europe, as well as to Boston and New
York. Boeing 737-200s make up the main part of Aer Lingus' fleet.
(P Hornfeck)

Rear: **Aloha Airlines** (AQ)
The company was formed on 9 June 1946 under the name of Trans-
Pacific Airlines. Non-scheduled operations commenced two months
later, followed by the introduction in 1949 of scheduled services over
an inter-island network linking the five main islands of Hawaii. Today
the company utilises its 737-200 fleet to offer high-frequency
services of over 100 flights a day, seven days a week, operating under
the name Aloha Airlines, which was adopted in 1959.
(Andy Clancey)

Title Page: **Boeing Commercial Airplane Company**
The Boeing 737-100 made its inaugural flight in April 1967, entering
commercial airline service with Lufthansa on 10 February 1968 and
with United Airlines later that year, The aircraft has sold well over
1000 examples and is set to become the world's best-selling
commercial jet aircraft.

Introduction

At the end of 1958 it was announced by Boeing that a design study would be made for a twin-engined "feeder" airliner to complete the family of Boeing passenger jets. The increase in air transport showed a ready market for such an aircraft to serve the smaller communities. In February 1965 the company decided to enter this field and officially announced that it was to build a twin-engined jetliner, the 737.

Two months passed and United Air Lines placed an order for 40 examples, supplementing the launch order for 20 from Lufthansa. With these orders the 737 was officially launched. The aircraft was specifically designed as a "short-haul" transport and replaced many DC-3s on routes of 400 km or less. One of the shortest routes operated by a 737 has been the route served by United Air Lines, flying the approximate distance of 50 km between Muskegan and Grand Rapids.

The initial variant of the 737 was the series -100, accommodating between 80 and 101 passengers with baggage. Power plants were Pratt and Whitney JT8D-7 engines. Other variants to the basic -100 series included the -100C, a convertible passenger/cargo version with a 2·21 × 3·66 m (87 × 144 in) forward cargo door and standard pallet loading system that is used on the larger Boeing transport jets. This type could carry six pallets totalling 12 700 kg (28 000 lb) in weight for a maximum of 480 km (300 miles). The -100E was an executive version, seating 25 persons in a custom-built interior over a maximum range of 5630 km (3500 miles), with reserves. An aero-medical variant, the -100M, was made available and accommodated up to 44 stretchers plus attendants.

The series 200 is basically similar to the initial model but is equipped with upgraded JT8D-9 engines and given an extended fuselage of 1·83 m (6 ft), accommodating between 88 and 113 passengers. A -200C convertible passenger/cargo version is available as well as a -200QC "quick-change" variant which allows faster conversion by use of palletised passenger seats and galleys as well as advanced cargo loading techniques.

In early 1980 work commenced on a new lengthened version, the -300 series. This airframe has an approximate 70 per cent commonality with the Advanced -200 series. The stretching of the fuselage allows extra passengers and freight to be accommodated. Two CFM56-3 turbofan engines complete the "new look" for this model, making it far quieter and more economical to run. Having received the go-ahead for production in March 1981 with orders for the type from US Air and Southwest Airlines, work commenced in mid-1982. The prototype 737-300 was rolled out ahead of schedule on 17 January 1984, making its first flight just over five weeks later. Certification was granted during November 1984 and it went into revenue service with Southwest Airlines on 7 December of that year.

I would like to extend my sincere thanks to Andy Clancey and Udo Schaefer, who were kind enough to send their transparencies for inclusion in this book. I would also like to thank Tony Carder and Brian Richards for their overall help in the preparation of this manuscript. It is to these people that I dedicate this book.

TABLE OF COMPARISONS			
	737-100	**737-200**	**737-300**
Max. accommodation	115	130	149
Wing span	28.35 m (93 ft 0 in)	28.35 m (93 ft 0 in)	28.88 m (94 ft 9 in)
Length	28.65 m (94 ft 0 in)	30.48 m (100 ft 0 in)	33.40 m (109 ft 7 in)
Height	11.28 m (37 ft 0 in)	11.28 m (37 ft 0 in)	11.13 m (36 ft 8 in)
Max. t/o weight	44 000 kg (97 000 lb)	52 390 kg (115 000 lb)	56 472 kg (124 500 lb)
Max. cruis. speed	925 km/h (575 mph)	927 km/h (576 mph)	899 km/h (558 mph)
Maximum range	3860 km (2400 miles)	4075 km (2530 miles)	4225 km (2625 miles)
Service ceiling	9145 m (30 000 ft)	9145 m (30 000 ft)	10 670 m (35 000 ft)

Opposite: **Air Pacific** (FJ)

Air Pacific, the national airline of Fiji, is a major Pacific regional carrier that serves several domestic points as well as over 13 foreign destinations. Established in 1951 as Fiji Airways, the airline commenced operations in September of that year using de Havilland Dragon Rapides. In 1957 Qantas, the Australian national airline, purchased the company and subsequently opened international routes on its behalf. By the late 1960s several Pacific islands had joined the Government of Fiji, Air New Zealand, BOAC and Qantas as major shareholders. In 1971 the airline changed its name to Air Pacific and by 1978 the Fiji Government had acquired all other airline shares to gain a majority holding. Today Air Pacific has a wide route network, with services to destinations that include Brisbane, Melbourne, Sydney, Auckland, Suva, Nadi, Noumea and Pago Pago. The company's sole 737-200 is seen here during a stopover in Auckland. *(P Bish)*

Air Algérie (AH)

The airline was formed in 1947 as Compagnie Générale des Transports Aériennes. In 1953, following a merger with Compagnie Air Transport, Air Algérie was formed. Ten years later the airline became the country's flag carrier, the government acquiring a majority shareholding of 51 per cent. Total nationalization took place in 1972. Two years later the airline took over operations of Société de Travail Aérienne. Today, Air Algérie scheduled operations connect the major cities of Algeria with over 35 foreign destinations across north, west and central Africa, the Middle East and Europe. The company's large fleet of 737-200s serves both domestic and regional routes, and operate as back-up equipment for the airline's two Airbus A310s on the prime route to Paris.

(Udo and Birgit Schaefer Collection)

4

Air Atlantis (TP)

The company was formed on 1 May 1985 by Dr Jose M S Gomes Motto, to take over and expand the charter services of TAP Air Portugal, with its base at Faro. A variety of aircraft are leased from the parent company, including Boeing 707, 727 and 737 types. A typical route flown by the company, is Faro-London (Gatwick). The airline operates its aircraft in a high-density all-economy class layout, thus flying its 727s with 131 seats instead of the normal 118, the 707s with 189 seats instead of 168 and the 737s with 133 instead of 130.
(Udo and Birgit Schaefer Collection)

Opposite: **Air Belgium** (AJ)

Air Belgium is a privately-owned charter company that flies international tour group services with Boeing 737-200 and -300 aircraft. The airline was established in May 1979 as Abelag Airways and commenced flight activities a month later. The majority of Air Belgium's tour group flights are operated between Brussels and holiday resorts in France, Greece, Italy, Yugoslavia, North Africa, Madeira, Spain and the Canary Islands. The airline's

737-200 is seen here at one of its Mediterranean destinations. *(DPR Marketing & Sales)*

Above: **Air Berlin USA** (ZF)

Air Berlin is an American company that has its base at Berlin's Tegel Airport. The company operates passenger charters with a Boeing 737-200 leased from Hapag Lloyd. The airline was formed in 1978 as a wholly-owned subsidiary of Lelco, an Oregon-based land and timber industry company. Charter

flights began in April 1979 with a flight to Palma de Mallorca. Between October 1980 and October 1981 scheduled passenger services were operated between Berlin, Brussels and Orlando in the USA, utilizing Boeing 707 aircraft. Today charters are flown to the Canary Islands, Balearic Islands, Mediterranean countries, the mainland of Spain, Portugal, Madeira, Austria and Israel, the latter having a regular service between Berlin and Tel Aviv. *(Udo and Birgit Schaefer Collection)*

Above: **Air Cal** (OC)

Air Cal was established in April 1966 as Air California. Flight operations commenced nine months later using Lockheed L-188 Electra aircraft, initially linking Orange County with San Francisco. Westgate-California Corporation assumed a controlling financial interest in 1977 and in 1981 Air California became Air Cal, complete with a new corporate logo and livery. The company is a major regional airline that operates scheduled passenger services to over 20 points in the western states of California, Nevada, Oregon and Washington. The carrier also flies charters throughout the United States and Mexico. Destinations include Burbank, Los Angeles, Oakland, Seattle and San Francisco. Air Cal's fleet consists of a mixture of McDonnell Douglas MD80, BAe 146 and 737 aircraft of both the -200 and -300 variants.
(Udo and Birgit Schaefer Collection)

Opposite: **Air Europe** (AE)

Air Europe is a major British airline that maintains international tour group flights. The carrier, a subsidiary of Intasun Leisure Group, was formed in 1978 as Inter European Airways. However, by the time of the company's first flight on 4 May 1979, the airline had changed to its present name. The primary areas of flight activity are the Mediterranean, the Canary Islands and the Middle East, as well as to holiday destinations within Europe. Air Europe has British flight bases at London (Gatwick) and Manchester, and flies from other UK airports that include Birmingham, Cardiff, East Midlands, Leeds and Luton. The airline operates an all-Boeing fleet, which is made up of 737-200s, as seen here, as well as 757-200 aircraft. The company in 1986 had orders outstanding for the 737-300 variant as well.
(Andy Clancey Collection)

Air France (AF)

Air France provides scheduled passenger and freight flights over one of the largest and most enduring networks in the world. The Government-controlled French national carrier links France with over 130 destinations worldwide, and in addition serves a limited domestic network centred upon the trunk route between Paris and Nice. The carrier was founded on 30 August 1933 following the takeover of Compagnie Générale Aéropostale by SCELA (formerly Air Orient, Air Union, CIDNA and SGTA), at this time operating a total of 259 aircraft of 32 types. With nationalization in 1945 and its legal establishment as national airline in 1948, Air France has evolved steadily to its present position as fourth among European carriers in terms of passenger traffic. Air France received the first of its Boeing 737s in 1982. However the company had used examples leased from Western Airlines in the mid-1970s for services between French Caribbean islands. *(Udo and Birgit Schaefer Collection)*

Air Liberia (NL)

Air Liberia was established as the country's national carrier in 1974 following the merger between Liberian National Airlines and Ducor Transport. From its base at Monrovia the company now utilizes a fleet of feeder and light aircraft on a comprehensive domestic route system in addition to regular regional charter flights. Air Liberia operated this single example of the 737-200C from 1978 until its sale to Lina Congo in 1982 on passenger and cargo services throughout Africa, many on behalf of the national government.
(DPR Marketing and Sales)

Above: **Air Madagascar** (MD)

Air Madagascar, formed jointly in 1961 by the government of the Malagasy Republic and Air France, has the distinction of operating one of the densest domestic airline route systems in Africa, covering more than 50 destinations. The airline's fleet of two 737-200s served the trunk routes within the island in addition to regional services to points in East Africa and the Indian Ocean, including Réunion, Nairobi and Moroni (Comoro Islands). In conjunction with Air Mauritius a route between Tananarive and Port Louis is also operated. *(DPR Marketing and Sales)*

Opposite: **Air Malta** (KM)

Air Malta was formed in March 1973 and, inaugurated services a year later over a Malta-Rome route. Control of the airline is held by the Maltese Government, which owns 96·4 per cent, Intercontinental Services LTD (3 per cent) and Cassar & Cooper Holdings (0·6 per cent). Air Malta Holdings include a 25 per cent shareholding in Med Avia, the Maltese-based oil-related charter company, plus investments in tour and travel, insurance and hotel companies. Scheduled flights using Boeing 737-200 and Boeing 720-047B aircraft link Luqa Airport with Amsterdam, Catania, Frankfurt, London, Lyon, Munich, Paris, Rome and Zurich in Europe, as well as Cairo and Tripoli in North Africa. Primary routes, however, are Malta-Rome and Malta-London, which have daily non-stop services. About one third of the airline's passenger traffic is carried on charter flights. Inclusive tour flights are carried out by Air Malta Charter using the parent company's aircraft as required. *(P Hornfeck)*

Below: **Air Nauru** (ON)

Air Nauru began operations in February 1970 with a service from Nauru to Brisbane, operated with leased Falcon executive jet equipment. Two years later the company obtained its own jet aircraft when it acquired a Fokker F-28 to increase capacity on its services. Boeing 727-100 and 737-200 aircraft were purchased in 1976 to supplement the fleet. Points served include Auckland, Noumea, Sydney, Spain, Nadi, Melbourne and Honiara.
(Udo and Birgit Schaefer Collection)

Opposite: **Air Sinai** (4D)

Air Sinai, The Egyptian regional airline, began operations in April 1982, the same month that Israel completed its final phase of negotiations for the return of the Sinai to Egypt. The company had succeeded Nefertiti Aviation as Egypt's flag carrier over the Cairo-Tel Aviv route. Today the company operates international services not only to Tel Aviv but also to Eilat. Destinations served include Al Areish, Ras An Naqb, Sharm el Sheikh, Mersa Matruh and Hurghada. The company's Boeing 737-200 is supplied on a lease basis from Egyptair, although the airline does own three Fokker F-27s.
(DPR Marketing & Sales)

Airways International – Cymru (AK)
The company was formed on 1 February 1984 by
Mr Tony Clemo of Red Dragon Travel. The Cardiff-
based airline initially used BAe 1-11 Series 300s on
inclusive tour flights throughout Europe, the
Mediterranean and North Africa, as well as
occasional services to the Middle East. A Boeing
737-200 has been leased from Guiness Peat Aviation since March 1985.
(Udo and Birgit Schaefer Collection)

16

Alaska Airlines (AS)

Alaska Airlines maintains scheduled passenger flights along an expanding network that now connects over 90 points. The Company can trace its history back to 1932, when McGee Airways began an Anchorage-Bristol Bay service. The airline was later merged into Star Airways, which eventually became Alaska Star Airlines. In 1944, following further consolidations, Alaska Airlines was formed.

Seven years later the airline opened up a Fairbanks-Anchorage-Seattle route. Today the airline's services include mainline flights within Alaska and to the Pacific Northwest and California, interchange flights across the US mainland, and regional and local commuter operations in western Alaska. The company operates a fleet of Boeing 727, 737-200 and McDonnell Douglas MD80 aircraft. *(Udo and Birgit Schaefer Collection)*

17

America West Airlines (HP)

America West maintains low-fare scheduled passenger services over an ever-expanding route system now linking more than 20 points in the southeast, California, midwest and western Canada. The carrier has a flight hub at Phoenix Sky Harbour Airport. The company was established in September 1981 and commenced scheduled flights two years later with Boeing 737 aircraft. Initial routes connected Phoenix with Los Angeles, Colorado Springs, Kansas City and Wichita. The airline today operates a fleet of Boeing 737s made up of -100 -200 and -300 variants. Points served include Burbank, Salt Lake City, Las Vegas, Ontario, Omaha, Palm Springs and San Jose. A 737-300 is seen here at Phoenix. *(S Ahn)*

Bahamasair (UP)

Bahamasair was established in June 1973 as the national airline of the Bahamas, when the country's government purchased two private carriers, Out Island Airways and Flamingo Airlines. Today the airline maintains scheduled passenger flights to over 20 points throughout the Bahamas, as well as to international destinations in the United States and to the Turks and Caicos Islands. Regular charter services are undertaken to many varied destinations, including a service between Nassau and New York (Newark) airport. Scheduled routeings include Nassau-Miami-Fort Lauderdale, Marsh Harbour-Treasure Bay-Miami, Nassau-Freeport-Atlanta, Nassau-West Palm Beach, with the company's primary service linking Nassau and Miami, flown by multiple Boeing 737-200 non-stop flights. *(Udo and Birgit Schaefer Collection)*

Braathens SAFE (BU)

Braathens SAFE is a privately-owned Norwegian airline that operates both scheduled domestic passenger flights and international tour and group charters. The airline was formed by ship owner Ludvig G Braathen in March 1946, with charter flights commencing a year later. Initial charter activities included flights to northern South America and the Far East, so creating the acronym SAFE in the airline's name. Today the company operates a fleet of Boeing 737-200 and Fokker F-28 aircraft and schedules over 130 daily departures systemwide along a route pattern concentrated in the south of Norway. Charter services cover Scandinavia and Europe, and extend to points in Africa and Asia.

(Udo and Birgit Schaefer Collection)

Britannia Airways (BY)

Britannia is one of the world's leading charter airlines, and is Britain's largest tour group carrier. The airline undertakes extensive international passenger and freight charters, as well as private and government contract services. The company also undertakes aircraft leasing activities with an all-jet fleet, the majority of which are Boeing 737-200s. Britannia is controlled by International Thomson, who also own the leading British tour company, Thomson Travel. The company began operations in May 1962 as Euravia Ltd, initially using Lockheed Constellation equipment. Today the bulk of the airline's flights are to destinations in Europe, the Mediterranean, West Africa, the Canary Islands, north and east Africa as well as to the Middle East. *(Udo and Birgit Schaefer Collection)*

Opposite: **British Airways** (BA)

The company is one of the world's leading airlines with a route network of over one million kilometres. British Airways was formed in April 1974 through a merger of British Overseas Airways Corporation and British European Airways. It was not however until 1977, when the European and Overseas divisions were finally merged, that the airline operated as a single entity. In 1976 British Airways jointly with Air France inaugurated the world's first supersonic passenger service using Concorde. These now link Washington, Miami and New York, as well as undertaking worldwide charters. The company's subsidiaries include the charter carrier British Airtours, as well as travel, aviation mainten-ance, flight support, catering, technical and training firms. The airline flies from all major points in the UK to Europe, the USA and Canada, South America, Africa, the Middle and Far East as well as Australia and New Zealand. Boeing 737-200 operations commenced in the late 1970s when the company leased three examples from the Dutch charter airline Transavia. *(A Meredith)*

Above: **Condor Flugdienst** (DF)

Condor, a wholly-owned subsidiary of Lufthansa German Airlines, is one of the world's largest charter air carriers, operating intercontinental tour and general passenger group flights with an all-jet fleet. The company was formed in 1955 as Deut-sche Flugdienst, with flight activities beginning a year later using Vickers Viking aircraft. In 1960 the carrier became fully owned by Lufthansa and in 1961 adopted its present title. Regular tour charter services are operated from Frankfurt, Dusseldorf, Munich and Stuttgart. Other German gateways include Hamburg, Bremen, Hanover and Saar-brücken. Condor operates package tour flights mainly to destinations in Spain, Greece, the Canary Islands, Italy, Portugal, Turkey, Tunisia, Morocco, Kenya, Thailand, Sri Lanka, the United States, Canada and Mexico. Condor's fleet is made up of Airbus A310, Boeing 727, 737-200, and DC-10 aircraft. *(Udo and Birgit Schaefer Collection)*

Continental Airlines (CO)

Continental Airlines provide scheduled low-fare jet services to nearly 50 domestic destinations, and to foreign airports in the South and Central Pacific, East Asia, Mexico, Canada and the United Kingdom. The company dates back to 1934, when Varney Special Lines began services. In 1936, following the purchase of a Denver-Pueblo route from Wyoming Air Service, the company moved its base from El Paso to Denver, and changed its name to Continental Airlines. In October 1982 the company merged with Texas International Airlines. Today the airline has a fleet that consists of Douglas DC-9, DC-10, Boeing 727 and Boeing 737-300, as well as McDonnell Douglas MD80 aircraft. The carrier has formed a subsidiary, Continental West, which operates a fleet of 737-300 aircraft.
(Udo and Birgit Schaefer Collection)

CP Air (CP)

CP Air, Canada's second largest airline, maintains scheduled passenger flights to over 17 points across Canada as well as to 14 points in the USA, South America, Europe, the Orient and the South Pacific. The airline is a division of Canadian Pacific, a diversified corporation that has additional holdings in trucking, rail, shipping, hotel and telecommunications systems. The airline operates a fleet of DC-10 and Boeing 737-200 aircraft. Points served include Calgary, Edmonton, Halifax, Vancouver, Amsterdam, London, Lisbon, Rome, Buenos Aires, Lima, Los Angeles and San Francisco.
(R Vandervord)

Opposite: **Dan-Air Services** (DA)

The company was founded on 21 May 1953 with a base at Southend as a subsidiary of Davies and Newman Holdings, the London ship brokers, from which its name is derived. Subsequent moves of headquarters were to Blackbushe in early 1955, and from there to London (Gatwick) Airport in 1960. Scheduled flights began in 1956 over a Blackbushe-Jersey line. The company acquired Scottish Airlines in 1961 and Skyways International 11 years later. Dan-Air has developed an extensive inclusive-tour charter operation network from most of the major UK cities and West Berlin. The airline also has an increasing scheduled route system that covers Britain, Ireland, France, Holland, Norway, Switzerland, West Germany and Austria. Dan-Air has operated Boeing 737s under a leasing system since the early 1980s with series-200 aircraft. A -300, the latest in the 737 range, was delivered in 1985 as G-SCUH.
(Boeing Commercial Airplane Company)

Below: **Delta Air Lines** (DL)

Delta Air Lines was established in 1924 as a crop-dusting company, with a base at Macan, Georgia. Four years later the airline changed its name to Delta Air Service, with scheduled passenger operations beginning in June 1929 over a Dallas-Shrevport-Monroe-Jackson route. On 1 May 1953 Delta was merged with Chicago and Southern Air Lines to create Delta-Chicago and Southern Airlines (a title which was only retained for two years). Today the company ranks as one of the largest airlines in the world and currently schedules well over 1000 daily jet flights to almost 100 cities throughout the USA, Canada, Bermuda, the Bahamas, Puerto Rico and Europe. A Delta Boeing 737-200 is seen here arriving at the airline's base airport of Atlanta, Georgia.
(Udo and Birgit Schaefer Collection)

Opposite: **Dragonair** (DE)

Dragonair is a newly formed Hong Kong-based operator that flies scheduled services to mainland China, as well as other destinations in southeast Asia. The airline operates a single Boeing 737-200 leased from the Irish leasing company, Guiness Peat Aviation. It was hoped that the carrier would commence "wide-bodied" services to London in the late 1980s.

(Udo and Birgit Schaefer Collection)

Eastern Provincial Airways (PV)

EPA was formed in 1949 flying mail, medical and survey services with Norseman aircraft from a base at St John's. Regular passenger flights began in 1955 between St John's, Gander and Deer Lake. Scheduled services were inaugurated in 1961 and by 1986 included Toronto, Montreal, Charlottetown, Halifax, Deer Lake, St John's, Wabush and Gander. EPA became a wholly-owned subsidiary of Newfoundland Capital Corporation following the latter's establishment in 1980. Two years later the airline created its Air Maritime division which operates BAe 748 aircraft. Eastern Provincial Airways and its subsidiary have been integrated into Canadian Pacific Airlines. A 737-200 in EPA's new colour scheme is seen here while on a visit to Fort Lauderdale's Hollywood Airport.

(Udo and Birgit Schaefer Collection)

Egyptair (MS)

Egyptair was established in 1932 as Misr Air Work, with a change of name to Misrair in 1949. In 1960 the company again changed its name, this time to United Arab Airlines, finally adopting the present title in 1971. Today the company operates scheduled flights over domestic routes in Egypt and international services in the Middle East, Africa, southern and eastern Asia, as well as Europe. The carrier is also a major Arab charter operator. Points served include Abu Simbel, Cairo, Luxor, Manila, Bangkok, Kuwait, Brussels, Geneva, London, Paris, Rome, Zurich, Dar Es Salaam and Nairobi. The airline's fleet is made up of Boeing 707, 737-200, 747, 767 and Airbus A300 aircraft.

(R Vandervord)

Hapag Lloyd Fluggesellschaft (HF)

Hapag Lloyd Flug, one of West Germany's largest charter airlines, operates an extensive programme of inclusive-tour passenger flights in addition to regular worldwide cargo services utilizing Airbus A300C4 Convertible equipment. The carrier is a subsidiary of the Hapag Lloyd Shipping Group and maintains its operations and engineering base at Hanover airport. The company was established in July 1972 and began flight activities eight months later. By 1978 the airline had gained financial control of Bavaria Germanair. The majority of the company's flights link West Germany with worldwide holiday destinations, extending throughout Europe to southern Asia, most notably Sri Lanka and the Maldives. Primary German gateways include Dusseldorf, Frankfurt, Hamburg, Hanover, Munich, Stuttgart, Bremen and Cologne/Bonn. A 737-200 is seen here at Dusseldorf.
(Udo and Birgit Schaefer Collection)

31

32

Hispania Lineas Aereas (HI)

Hispania is a charter airline that operates international passenger inclusive-tour flights, primarily between Spain and destinations in Europe and Africa. The company, formed by a number of ex-TAE employees, began operations on 28 April 1983 with a round trip charter between Palma and Seville using an ex-Transeuropa Caravelle. The airline has its base and headquarters at Palma de Mallorca in the Balearic Islands. Hispania's colourful livery is seen here on a 737-200 on a flight to London (Gatwick). *(R Vandervord)*

Lauda Air Luftfahrtgesellachaft (LD)
The company was formed in May 1979 by former world champion racing driver Nikki Lauda. The airline is a charter carrier that undertakes executive air taxi and leasing services. In 1985 Lauda Air expanded into passenger charter flights by operating services for a number of tour operators to destinations in the Mediterranean and to Southern France. In addition to this, panoramic round-trips and special flights to racing and other events are operated. Boeing 737-200 operations began in 1985 when an aircraft was leased from Transavia, the Dutch charter operator.
(Udo and Birgit Schaefer Collection)

Lufthansa German Airlines (LH)

Lufthansa, the West German national airline, operates scheduled services over an intercontinental route system that connects nine domestic and over 118 foreign points. The company's history can be traced back as far as 1926, when Deutsche Luft Hansa was formed through a merger of Aero Lloyd and Junkers Luftverkehr. At the start of World War 2 the company was disolved. However, in 1953 a new West German national airline known as Luftag was formed. The carrier was renamed Deutsche Lufthansa a year later with international services commencing shortly afterwards. The company was the first airline to place an order for the Boeing 737-100 and today operates both the -200 and -300 variants. Lufthansa's international flights link West Germany with the rest of Europe, North and South America, the Caribbean, Africa, the Middle East, Asia and Australia. *(Udo and Birgit Schaefer Collection)*

Luxair (LG)

Luxembourg's national carrier Luxair, was formed in 1962 to provide regional passenger schedule and charter services. Operations commenced with a service to Paris, a route now operated daily along with those to Frankfurt, Zurich and London (Heathrow), using Fokker F-27 or 737-200 equipment. Regular inclusive-tour services are operated to Mediterranean resorts in addition to seasonal scheduled services to Palma de Mallorca. On behalf of the charter company Luxavia, Luxair operates regular international passenger flights to Johannesburg using Airbus A300 aircraft.
(DPR Marketing and Sales)

Opposite: **Maersk Air** (DM)

Maersk Air was established in 1969 as a subsidiary of the A P Moller Shipping Company (Maersk Line) and started flight operations in December of that year. By early 1970 the airline had absorbed the Danish domestic airline Falckair. The following year the carrier had joined both SAS and Cimber Air to form Danair, Denmark's main internal air carrier. Today Maersk Air is a major Danish airline that maintains domestic scheduled flights for Danair, and provides a scheduled service between Billund and Southend, United Kingdom, as well as exten-sive charter and contract operations. The airline maintains a fleet of de Havilland Dash 7, BAe 125, Bell 212, Super Puma and Boeing 737 aircraft, the latter type being a mixture of both -200 and -300 series. *(A J Mercer)*

Above: **Markair** (BF)

Markair is a rapidly expanding airline that provides scheduled passenger and cargo services through-out the state of Alaska. The carrier was formed in 1947 and was known as Interior Airways until its name change in 1972 to Alaska International Airlines. In September 1980 the company acquired Great Northern Airlines, and with the inauguration of passenger Boeing 737-200 flights on 1 March 1984 changed to its present name. Initial routes linked Anchorage with Fairbanks, Barrow and Bethal. Also during the same year, Markair greatly increased its presence in Alaska by acquiring most of the airport facilities of Wien Airlines. Points served include; Anchorage, Aniak, Barrow, Bethel, Dillingham, Fairbanks, Galena, King Salmon, Kodiak, Kotzebue, Prudhoe Bay and St Mary's. *(Udo and Birgit Schaefer Collection)*

Above: Midway Airlines (ML)
The company began operations on 1 November 1979 from Chicago/Midway to Cleveland, Detroit and Kansas City. Midway has the honour of being the first major new US carrier to begin operations following passage in 1978 of the Airline Deregulation Act. The airline's route network covers Chicago (Midway), Cleveland, Dallas, Detroit, Kansas City, Milwaukee, Minneapolis St Paul, Newark, New York (La Guardia), Philadelphia, Topeka, Washington DC and White Plains. In 1984 the company gained court approval to establish a subsidiary, Midway Express, to resume the oper-

ations of the bankrupt Air Florida. Utilizing that company's former aircraft and employees, Midway Express made its first flight in October of that year, between Chicago and Miami, using Boeing 737-200s. *(Udo and Birgit Schaefer Collection)*

Opposite: Monarch Airlines (OM)
Monarch is a leading UK charter carrier that operates holiday tour and general passenger group flights. The airline also undertakes aircraft leasing services and performs engineering and mainten-ance work via an associate company, Monarch Aircraft Leasing. The carrier is owned by Cosmos, a

major British package tour operator for which the airline maintains extensive flight services. Monarch was established in June 1967 and commenced its first revenue flight ten months later with a service between Luton and Madrid using Bristol Britannia turboprop aircraft. The airline's main destinations are in Europe, the Mediterranean, North Africa, the Canary Islands as well as the Middle East. Primary UK flight gateways are Luton, Manchester, London (Gatwick), Birmingham and Glasgow. The airline operates a fleet of Boeing 757 and 737 aircraft. The latter type is made up of -200 and -300 variants. *(Andy Clancey Collection)*

Opposite: **New York Air** (NY)

New York Air was formed in 1980 to compete directly with Eastern Airlines and its ''Air Shuttle'' service between New York and Washington. The airline provides scheduled passenger services throughout the northeast USA, extending to the midwest, southeast and Florida. The carrier is a subsidiary of Texas Air, a Houston-based corporation which owns amongst other companies Continental Airlines. New York Air operates a fleet of 737-300, McDonnell Douglas MD80 and Douglas DC-9-30 aircraft. Points served include Boston, Detroit, Newark, New Orleans, Orlando, Rochester, Tampa and Washington DC (National and Dulles airports). *(N Chalcraft)*

Above: **Nordair Ltd** (ND)

Nordair was formed in 1957 by the merger of Boreal Airways and Mont Lauier Aviation. Three years later the company absorbed Wheeler Airlines, a major operator in the Hudson Bay region of Quebec and in the Canadian Arctic. Today the airline is a prominent Canadian regional operator, flying scheduled flights in Quebec, Ontario, Manitoba and the Northwest Territories, as well as to the United States. The carrier also undertakes cargo and charter contract services which include regular tour group Boeing 737-200 flights between Canada and Florida. Additional charters include those operated on behalf of the Canadian Government and the United States Air Force. Nordair provides much vital air transportation between southern Canada and the remote communities of northern Quebec and the Canadian Arctic.
(Udo and Birgit Schaefer Collection)

41

Olympic Airways (OA)

The government-owned Greek national airline Olympic Airways was formed in April 1957 when Aristotle Onassis took over TAE Greek National Airlines. The company maintains scheduled flights over an extensive domestic network as well as international routes within Europe and to the Middle East, Africa, southern Asia, Australia and North America. In 1974 Onassis relinquished his holding in the company and ownership was taken over by the government on 1 January 1975. Olympic has its major flight hub at Athens but operates additional services from other Greek airports such as Salonika, Corfu, Heraklion, Rhodes and Iona. The company's fleet is made up of Boeing 707, 727, 737-200, and Airbus A300 aircraft. *(R Vandervord)*

Orion Airways (KG)

Orion Airways is an expanding British airline which is primarily engaged in the transport of inclusive-tour groups over holiday routes in Europe, North Africa and the Canary Islands, The carrier was formed in 1980 by Horizon Travel, a large British package holiday operator. The first flight was operated in March of that year with a charter to Pisa from its base at East Midlands. Primary UK gateway airports are East Midlands, London (Gatwick), Birmingham and Manchester. Other regional airports served include Bristol, Cardiff, Bournemouth, Edinburgh, Glasgow, Leeds/Bradford, Luton and Newcastle. Orion's all-737 fleet consists of both -200 and -300 variants. An example of the former type is seen here at Cologne/Bonn airport.
(Udo and Birgit Schaefer Collection)

43

Pacific Western Airlines (PW)

Pacific Western, Canada's third largest commercial airline, maintains a scheduled route network that encompasses the entire western part of the country, with routes that reach as far east as Toronto. The company was established in 1946 as Central British Columbia Airways, the present title being adopted in 1953. A major expansion took place in 1959 when the airline took over various routes of Canadian Pacific Airlines in Alberta and the Northwest Territories. Further consolidation followed with the acquisition of BC Airlines in 1970. In 1978 PWA purchased Transair Ltd, a major central Canadian regional carrier. Today, in addition to maintaining an extensive network of scheduled services, the airline also undertakes international passenger group charters. Flights are regularly made to Mexico, the Caribbean and sunbelt cities in the United States. A 737-200 is seen here at Luton International Airport, just prior to its delivery to Canada. *(A J Mercer)*

Pakistan International Airways (PK)

PIA provides a scheduled route network throughout Pakistan, as well as to destinations in southern and eastern Asia, the Middle East, North and East Africa, Europe and the United States. The national flag carrier was established by the government in 1951 and began flight operations four years later, when it acquired and replaced the privately-controlled Orient Airways (which itself was established in 1946). The airline's international route system has a main hub at Karachi, with additional Pakistani gateways at Islamabad/Rawalpindi, Lahore and Peshawar. The company operates six Boeing 737-300. Two examples are seen here at London (Heathrow) Airport while on delivery to Karachi. *(R Vandervord)*

Above: **Pan American World Airways** (PA)

Pam Am, one of the leading US carriers, was formed in 1927, having inaugurated a regular mail service between Key West and Havana. In the following years the airline became established as perhaps the most recognisable name and image in commercial air transport history. By the 1930s routes were opened to Mexico, the Caribbean, Central and South America. Today the airline provides a scheduled route network that links over 90 points in 50 countries around the world. Pan Am has its headquarters in New York and bases its aircraft at New York (JFK), Miami, Los Angeles, San Francisco, London (Heathrow), Frankfurt, Berlin and Tokyo. Pan Am's fleet of 737-200s, based in Europe for internal German and regional services, are leased from various carriers worldwide. *(P Hornfeck)*

Opposite: **People Express Airlines** (PE)

People Express was established on 7 April 1980 and began scheduled flights a year later from Newark to Buffalo, Columbus and Norfolk. Since then the company has expanded dramatically and now provides scheduled jet services to over 40 points across the United States and Europe, operating from a major flight hub at Newark International Airport. People Express have emerged from US deregulation as one of the most successful companies in America. In May 1983 the carrier commenced services to London (Gatwick). Points served include Baltimore, Chicago, Houston, Orlando, Sarasota, Oakland and Rochester. People's domestic network is operated by Boeing 727 and 737 aircraft, the latter type made up of -100 and -200 variants. *(N Chalcraft)*

Below: **Piedmont Airlines** (PI)
This company is a rapidly expanding domestic carrier that operates scheduled jet services to over 70 points across the United States. Piedmont began operations as an aircraft sales and service company in 1940, with an airline division being established in 1948. Initial routeings were between Wilmington in North Carolina and Cincinnati in Ohio. Today the company has a network concentrating in the Mideast, Northeast, Southeast and Midwest, with extensions to Colorado, Texas and California. Piedmont ranks now amongst the top ten airlines in the US in terms of passenger traffic and can boast the fact that it operates the world's largest fleet of Boeing 737s, of which it flies both the -200 and -300 series. The airline is the subsidiary of Piedmont Aviation, a corporation which also operates general and business aviation sales, service and maintenance companies. In 1983 Piedmont finalized an agreement to acquire fully Henson Airlines by October 1987. The well-established regional operator would retain its own identity, although their aircraft were to be painted into the "PI" colour scheme. *(Udo and Birgit Schaefer)*

Opposite: **Rotterdam Airlines**
The privately-owned Rotterdam Airlines was established in 1977. Scheduled services, however, did not commence until 1983, when a leased Boeing 737-200 was used over a Rotterdam-London (Gatwick) route, subsequently complimented by a Rotterdam-Innsbruck service. Unfortunately, due to serious financial difficulties, the airline ceased operations in 1984.
(Udo and Birgit Schaefer collection)

Royal Air Maroc (AT)

The airline was established in 1953 as CCTA following a merger between Air Atlas and Air Maroc, becoming the national flag carrier of the newly-independent Morocco in June 1957. RAM is controlled by the Moroccan Government, which owns a 90 per cent shareholding, Air France (5·5 per cent), and private interests. The company itself has interests in Royal Air Inter and occasionally leases 737 equipment to this operator of short-haul domestic services. Royal Air Maroc's scheduled flights connect major points within the kingdom to an extensive international route network. Points served include Rio de Janeiro, Montreal, New York, London (Heathrow), Athens, Rome, Frankfurt, Geneva and Madrid. The airline's fleet consists of Boeing 707, 727, 737-200 and 747 aircraft, with two fuel-efficient 757-200s on order for delivery in the fourth quarter of 1986.
(DPR Marketing and Sales)

Royal Brunei Airways Ltd (BI)

Royal Brunei Airlines was formed on 18 November 1983 as the government-owned national airline of the Sultanate of Brunei. The company currently provides regional jet services in south-east Asia from its base at Bandar Seri Begawan. Singapore is served by nearly 20 direct flights per week. In addition regular schedules are maintained to Hong Kong, Bangkok, Manila, Kuala Lumpur, Kuching and Kota Kina Balu, in Malaysia. There are also services to Darwin in Australia and from 1987 the airline planned to begin flights to Europe. Until the end of May 1986 the airline operated an all 737-200 fleet. However, Royal Brunei has since accepted delivery of its first 757-200ER.

(Udo and Birgit Schaefer Collection)

Above: **Sabena – Belgian World Airlines**
(SN)
Established in 1923, Sabena is one of the world's
oldest airlines, with the company's first flight com-
ing a year later over a Brussels-Strasbourg route.
Today the airline provides regular scheduled flights
to over 70 points worldwide. Major destinations
include Anchorage, Boston, Chicago, Bombay,
Tokyo, Paris, London (Heathrow), Jeddah and
former Belgian colonial centres in Brazzaville, Kin-
shasa, Kigali and Bujumbura. The central flag
carrier is owned by the government and various
private interests. Sabena itself has a majority shareholding in the charter carrier, Sobelair. The company operates a mixed fleet of Boeing 737-200,
747, Airbus A310, and DC-10 equipment. In the
near future Sabena/Sobelair expect to take delivery
of a number of 737-300 aircraft.
(Udo and Birgit Schaefer Collection)

Opposite: **Sahsa Honduras Airlines** (SH)
Sahsa, a major Central American airline, operates
scheduled flights over domestic routes in Honduras, as well as over international routes to Belize,
Costa Rica, El Salvador, Guatemala, Nicaragua,
Panama, and the United States. The company was
established in 1944 and commenced flight operations a year later. It is owned by private Honduran
interests (62%) and TAN Airlines (38%). Sahsa has
a majority shareholding in the Honduran domestic
airline Anhsa, with which it coordinates certain
local services. The airline's fleet consists of Boeing
727, 737-200 and Lockheed L188 Electra aircraft.
(Udo and Birgit Schaefer Collection)

(Udo and Birgit Schaefer Collection)

Opposite: **Southwest Airlines** (WN)
Southwest Airlines was formed in 1967, but due to legal battles with other well established companies the airline did not make its first flight until June 1971. The inaugural services were from Dallas (Love Field) to Houston and San Antonio. Now the carrier is well established amongst US airlines, with its primary route and traffic hubs at Dallas (Love Field), Houston Hobby Airport and Phoenix. Since US deregulation the company has grown rapidly and serves destinations that include Austin, Chicago, Denver, El Paso, Kansas City, Las Vegas,

San Francisco, and Tulsa. Southwest operates a fleet of Boeing 727 and 737 aircraft, the latter made up of -200 and -300 variants.
(Udo and Birgit Schaefer Collection)

Above: **Spantax Airlines** (BX)
Spantax, one of Spain's largest charter airlines, operates international inclusive-tour services to and from holiday destinations on the mainland of Spain, the Balearic Islands and the Canary Islands. The airline has its headquarters in Madrid, and bases its aircraft at Madrid, Palma de Mallorca and

Las Palmas. Spantax, which is privately-owned, was established as Spanish Air Taxis in October 1959, with services beginning two months later. In 1967 the airline acquired its first jet equipment in the form of Convair 990s. The company undertakes extensive tour group holiday charters throughout Europe, North Africa, North and South America. Regular transatlantic services include Madrid-New York with DC-10 aircraft. A Spantax 737-200 is seen here at Dusseldorf.
(Udo and Birgit Schaefer Collection)

TACA International Airlines (TA)

TACA was established in November 1939 as TACA El Salvador, and was one of a number of airlines that formed a group covering Central and South America. Today the privately-owned Salvadorian national airline maintains scheduled flights between El Salvador and Belize, Costa Rica, Guatemala, Honduras, Mexico, Nicaragua, Panama and the United States. the airline's fleet consists of Boeing 737-200, Boeing 767 and BAe 1-11 series 400 aircraft. *(DPR Marketing and Sales)*

TAP – Air Portugal (TP)

TAP, the Portuguese national airline, operates scheduled flights to over 40 points in Portugal, Western Europe, Madeira, the Azores, the Canary Islands, Africa, North and South America. The company was formed by the Portuguese government on 22 September 1944, with an initial flight between Lisbon and Madrid two years later. In 1946 the airline inaugurated an African route to Luanda and Lourenço Marques (now Maputo). Today TAP has its main flight hub at Lisbon, and maintains services from Oporto and Faro in mainland Portugal, Terceira in the Azores, and Funchal in Madeira. The company operates a fleet of Lockheed L1011 Tristar, Boeing 707, 727 and 737-200 aircraft. *(Udo and Birgit Schaefer Collection)*

57

Opposite: **Thai Airways** (TH)
Thai Airways was formed on 1 March 1947 by the government as the result of a merger between Siamese Airways, Aerial Transport Company of Siam, and Pacific Overseas Airlines (Siam). This new company took over the domestic and regional route systems of these carriers. In 1959, however, SAS signed an agreement to form Thai Airways International, which was designed to become the country's flag carrier and to take over international services. Today Thai operates an extensive domestic network radiating from its Bangkok base to over 20 other towns in Thailand as well as to Hanoi (Vietnam), Vientiane (Laos), Kuala Lumpur and Penang (Malaysia). Regional destinations served include Phitsanuloke, Lampang, Phrae, Chiang Mai, Mae Hongson, Chiang Rai, Nan, Khon Kaen, Udon, Loei, Nakhon Phanom, Uban Ratchathani, Phuket, Surat Thani, Trang, Hat Yai, and Pattani. Boeing 737-200 HS-TBC is seen here on arrival at Phuket. *(Udo and Birgit Schaefer Collection)*

Above: **Transavia Holland** (HV)
The company was formed in 1965 as Transavia (Limburg) NV, flying charters and inclusive-tour operations a year later. The first service was to fly the Dutch Dancing Theatre to Naples, using a Douglas DC-6, that type making up the airline's fleet. These aircraft were quickly replaced with jet equipment so that the company could increase its share of the Dutch holiday market. Transavia is a wholly-owned subsidiary of the Royal Nedlloyd Group, which also has a 49 per cent holding in Martinair Holland. The majority of the airline's commercial passenger charter activity involves inclusive-tour group transport to and from holiday areas in the Mediterranean, southern and western Europe, North Africa and the Canary Islands. The company commenced Boeing 737 operations in 1974 with series -200 aircraft. Today, with its new corporate identity, Transavia operates a mixed fleet of both -200 and -300 versions of the 737. *(Udo and Birgit Schaefer Collection)*

59

United Airlines (UA)

The western world's largest airline can trace its history back to 1931, when United Airlines was established through a merger of four companies, Varney Air Lines, National Air Transport, Pacific Air Transport and Boeing Air Transport. Today the company operates scheduled services to all states throughout the USA, and operates to foreign destinations in Canada, Mexico, the Bahamas and the Orient, as well as having just taken over Pan Am's Pacific routes. The airline is a main subsidiary of UAL Inc, which owns Western Hotels, GAB Business Services and Hawaii's Mauna Kea Properties. United has three primary system traffic and route hubs: Chicago O'Hare Airport, Denver's Stapleton Airport and San Francisco International Airport. Other major traffic centres include Los Angeles, Seattle and honolulu. The airline has either operated or was to operate all three main 737 types. *(Udo and Birgit Schaefer Collection)*

US Air (AL)

The airline was formed as All American Aviation in 1937 and commenced air mail flights in Pennsylvania and West Virginia two years later. In 1949, having changed its name to All American Airways, the carrier began scheduled passenger flights with DC-3 aircraft. The company again had a name change in 1953, appearing as Allegheny Airlines. Prior to US airline deregulation, the company grew very rapidly through various mergers. Lake Central Airlines was absorbed in 1968, which extended routes into the eastern Midwest, while Mohawk Airlines was purchased in 1972, extending and fortifying routes throughout the Northeast. This made Allegheny the country's premier regional airline and to establish an awareness of the company's wide network the present title was adopted in October 1979. Today US Air flies to most points in America as well as Montreal and Toronto in Canada. The company has since 1967 operated a commuter system under the banner "Allegheny Commuter". The airline operates a fleet of BAe 1-11, DC-9, 727 and 737-200 and 737-300 aircraft. *(Udo and Birgit Schaefer Collection)*

Below: VASP Brazilian Airlines (VP)
VASP was established in 1933 and has been owned since 1936 by the state of São Paulo. The airline undertakes extensive scheduled jet services to more than 30 points throughout Brazil and carries more domestic passengers than any other airline. VASP ranks as the country's second largest operator in overall size behind Varig. Points served include Aracaju, Belém, Brasilia, Fortaleza, Manaus, Rio de Janeiro and São Luis. The company is also responsible for certain "Ponte Aerea" shuttle flights that frequently link São Paulo Congenhas Airport with Rio de Janeiro Santos Dumont Airport. Services are provided with Varig L188 Electras. VASP's fleet is made up of Airbus A300, Boeing 727 and 737-200 aircraft. *(Andy Clancey Collection)*

Opposite: **Western Airlines** (WA)
Western is a major airline that provides scheduled passenger services covering a network of over 50 cities across the USA and nine destinations in Mexico and Canada. The airline was formed in 1925 with air mail flights between Los Angeles and Salt Lake City, via Las Vegas. Operations were expanded in 1944 to cross the Rocky Mountains. Today the airline serves locations which include Calgary, Edmonton and Vancouver in Canada, Acapulco, Guadalajara, Mexico City in Mexico and Anchorage, Burbank, Chicago, Houston, San Jose and Tulsa in the USA. Western operates a fleet of Douglas DC-10, Boeing 727 and 737 aircraft, the latter type comprising a mixture of -200 and -300 versions. *(Udo and Birgit Schaefer Collection)*

Wien Airlines (WC)

The company was established as Wien Consolidated Airlines in 1968, following a merger of Wien Alaska Airlines and Northern Consolidated Airlines. In 1973 the company changed its title to Wien Air Alaska. With US airline deregulation in 1978 vast changes were made to the company. Until then operations had been confined to the state of Alaska, where jet flights connected major cities and towns, and where additional services to over 100 villages were coordinated through service agreements with numerous commuter carriers. Due to financial difficulties Wien ceased operations on 28 November 1984 and filed for chapter 11 bankruptcy. Destinations that were served included Anchorage, Fairbanks, Kodiak, Oakland, Reno, Seattle and Portland. Until operations ceased Wien operated a fleet that included Boeing 727 and 737-200 aircraft.

(Udo and Birgit Schaefer Collection)